AF614674

# AN INTRODUCTION TO BUILDING MECHANICAL SYSTEMS

by

Tom Dontigny

T.D. AIR BALANCE SERVICE
261 John St North
Arnprior, Ontario
K7S 2P3
Phone : 613 623 2755
Fax: 613 623 9352
E mail: tdontig@sympatico.ca

Bloomington, IN   Milton Keynes, UK

authorHOUSE®

*AuthorHouse™*
*1663 Liberty Drive, Suite 200*
*Bloomington, IN 47403*
*www.authorhouse.com*
*Phone: 1-800-839-8640*

*AuthorHouse™ UK Ltd.*
*500 Avebury Boulevard*
*Central Milton Keynes, MK9 2BE*
*www.authorhouse.co.uk*
*Phone: 08001974150*



*First published by AuthorHouse 8/16/2006*

*ISBN: 1-4259-4857-X (sc)*

*Printed in the United States of America*
*Bloomington, Indiana*

*This book is printed on acid-free paper.*

# INTRODUCTION

The purpose of this **reference, training manual** is to provide an overview of building systems for both operators and administrators. It is presented in **easily understood layman terms**. It is intended that this manual can be used as a training tool and as a reference source for building operators, operations managers, property managers and tenant service personnel.

The manual has been divided into sections, with each section covering a particular portion of the mechanical systems. Each section outlines in general the characteristics of operation of the equipment. Comments regarding the interaction of components and systems, and their purpose for being provided as part of the building mechancial equipment package are included.

It is respectfully requested that this manual not be copied by any means without permission in writing from T.D. Air Balance Service.

# INDEX

# AIR SYSTEMS

The types of air systems and their operating components may consist of:

Terminal reheat constant air volume
Single duct variable air volume
Single duct constant air volume
Multizone constant air volume
Dual duct constant air volume
Single duct constant dump bypass air volume
Perimeter induction air system
Outside makeup air system
Fan coil
Parkade ventilation air system
Exhaust fans

Air system components

Thermostats
Pneumatic control air

## Terminal Reheat Constant Air Volume

The terminal reheat constant air volume system supplies a constant volume of air at a constant temperature through a single run of ductwork. This run of ductwork provides air to in duct heating coils serving individual areas or zones in the building. The air temperature is selected low enough to cool the highest heat gain area, usually at 58 to 60 deg F. The thermostatically controlled reheat coils heat the air delivered to areas which do not require the full cooling capacity of the air.

Room thermostats or the building automation system (BAS) through wall sensors modulate the reheat coil control valves to control the volume of heating water passing through the reheat coil to regulate the temperature of the air entering the area.

## Single Duct Variable Air Volume

The single duct variable air volume system supplies a varying quantity of air at a constant pressure and temperature in a single run of ductwork through variable air volume boxes to the air diffusers throughout the building. The volume of air delivered by the supply fan may be controlled by inlet control vanes, variable fan speed, variable fan blade pitch, outlet control dampers, which maintain a constant supply duct static pressure.

The pressure should be sufficient to maintain a minimum operating pressure at all variable volume control boxes. Operating at higher duct static pressure than neccessary wastes energy and results in higher noise levels at the VAV boxes due to throttling.

A wall mounted thermostat or sensor through the building automation system regulates the volume of air allowed to enter the controlled area through one or more ceiling mounted variable air volume boxes.

The return fan which draws air back from the space provides the return link in the air circulation for the zone.

## Single Duct Constant Air Volume

The single duct constant air volume system supplies a constant quantity of air at a varied temperature in a single run of ductwork to the air diffusers throughout the controlled area.

A controller or zone thermostat, sensing return, supply, or zone air temperature modulates the mixed air dampers, and heating and cooling coils to regulate the temperature of the air entering the area.

The return fan which draws air back from the space provides the return link in the air circulation for the zone.

## Multizone Air System

The multizone air system supplies air at a constant volume and varied temperature to a number of single duct runs which originate at the multizone unit. Each run of ductwork supplies air to diffusers in a specific zone.

Zone thermostats sensing the space temperature, operate a set of mixing dampers located at the main unit which regulate the mixed air temperature which is supplied to each zone. There is a hot deck and a cold deck.

The return fan which draws air back from all the zones provides the return link in the air circulation for the zones.

## Dual Duct Constant Volume Air System

The dual duct constant volume air system supplies air in two seperate runs of ductwork to various mixing boxes throughout the building. One air supply duct is called the hot deck and the other the cold deck with take offs to each mixing box located in the ceiling space.

Zone thermostats sensing the space temperature, operate a damper positioner in the mixing box or boxes to regulate the temperature of the mixed air supplied through the diffusers to the controlled area.

The return fan which draws air back from all the zones provides the return link in the air circulation for the zones.

## Single Duct Constant Volume Dump Bypass Air Volume

The single duct constant air volume system supplies a constant volume of air at a constant pressure and temperature in a single run of duct through VAV boxes to the ceiling mounted air diffusers throughout the building. The volume of air delivered by the supply fan is set at a constant volume.

The static pressure should be sufficient to maintain a minimum operating pressure at all VAV boxes. Operating at higher duct static pressure than necessary wastes energy and results in higher noise levels at the VAV boxes due to throttling.

Wall mounted thermostats sensing zone temperature regulate the volume of air allowed to enter the controlled area through the above suspended ceiling VAV boxes. On a call for cooling the VAV boxes deliver an increased air flow to the zone. Full cooling is the maximum air flow that the VAV box is set to deliver. Minimum cooling is the minimum air flow that the VAV is set to deliver. Minimum air flow is usually 25% of maximum air flow and ensures there is ventilating air available for the occupants during the heating season. Some rooms such as conference or interior training rooms or VAV boxes with electric reheat coils usually require a higher minimum air flow to ensure safe operating conditions.

When VAV boxes are in the minimum position air is bypassed to the above suspended ceiling space. Each VAV box is equipped with a bypass balancing damper at the top. This balancing damper should be set ensuring that too much air does not bypass to the ceiling space therefore "robbing" other VAV boxes of their required air.

## Perimeter Induction Air System

The perimeter induction air system is a constant volume system which supplies a constant quantity of primary supply air at a constant temperature to perimeter induction units.

An air system controller sensing supply air temperature modulates the mixed air dampers, preheat coil, heating and cooling coil control valves to regulate the temperature of the primary supply air.

The return fan draws air back from the floor space.

Induction units can be used for cooling only or cooling and heating purposes. During summer months chilled water is used in the coil and during winter months it is switched over to heating water.

Each perimeter induction unit has an adjustable air volume control damper. The primary air nozzles induce room air in through the front grille and through the coil then discharges at the top. This room air mixes with the primary air and maintains perimeter space temperature.

The induction unit heating and cooling function can be controlled by self contained controls or wall mounted thermostats.

## Outside Makeup Air System

The outside makeup air system supplies a constant volume of air at a constant or varied temperature in a single run of ductwork to make up a predetermined quantity of exhaust air.

A discharge controller or wall mounted thermostat which senses supply zone air temperature controls the heating section to regulate the temperature of the air entering the space. The heating section may be gas fired or glycol.

## Fan Coil

The fan coil supplies a constant volume of air at a constant temperature to the area served. The air may be full return air or a combination of return and outside air.

If operating on return air only the fan coil will have a chilled water cooling coil or evaporator coil with self contained DX cooling.

The blower fan can run continuously or be cycled on and off on a demand for heating or cooling.

A wall mounted zone thermostat cycles the blower, heating, cooling section on and off to maintain the serviced area at the set point of the thermostat.

## Parkade Ventilation Air System

The parkade ventilation system is necessary to maintain safe and acceptable carbon monoxide levels throughout the parkade. A sufficient amount of outside air is supplied to the parkade to replace the carbon monoxide laden air removed by the parkade exhaust system. The carbon monoxide is produced by the exhaust fumes of the vehicles entering and leaving.

It is not acceptable to let the carbon monoxide levels in the parkade to be in excess of 50 parts per million (p.p.m).

## Exhaust Fans

The exhaust fans simply remove a fixed amount of air while the supply fans provide a sufficient amount of outdoor air for adequate ventilation and pressurization of the building or garage.

The combined exhaust air and return air volumes should be less than the supply air volume. This surplus of supply air keeps the building at a slightly positive pressure. This inhibits infilteration of outside air through doors, windows and the building structure. This unwanted air infiltration results in increased energy consumption, uncomfortable temperature conditions and possible freeze up.

As a general rule washroom and storage area exhaust fans should be shut down at night.

## Air Systems Components

Air systems can be comprised of the following components;

Return Air Fan

Mixed Air Section With Minimum Fresh Air Damper
Filter Section
Preheat Coil
Preheat Coil Circulating Pump
Heating Coil
Heating Coil Circulating Pump
Cooling Coil
Cooling Coil Circulating Pump
Direct Expansion Cooling Unit (Evaporator)
Packaged Refrigeration Unit (Compressor and Condensor)
Humidifying Section
Supply Air Fan
Wall Mounted Thermostats
Pneumatic Control Air Compressor

## Return Air Fan

The return air fan draws air back from the building and delivers it to the mixed air section of the air system. The fan if equal in capacity to the supply fan air volume will exhaust some air through the exhaust dampers and deliver the rest to the supply fan suction side. If the fan is designed to return less air than the supply air fan delivers no air will go to exhaust when the mixed air dampers are at minimum outside air position.

The total of building return air plus building exhaust air must not exceed building supply air. If so, the building will be operating at a negative pressure relative to outside. This will allow for unconditioned outside air to enter the building through the building envelope. This is undesireable when very cold or hot outside.

On start up the return fan should start before the supply fan and on shutdown the supply fan should stop before the return fan.

The fan may be constant air volume or variable air volume depending on system design.

## Mixed Air Section With Minimum Outside Air Damper

The mixed air section consists of exhaust, return and outside air dampers. There may be a separate minimum outside air damper or the minimum could be a preset position of the outside air damper.

The mixed air section is usually controlled by sensing mixed air temperature and modulating the dampers to maintain a 60F air temperature. On an increase in mixed air temperature the outside air dampers open, the return air dampers close and the exhaust air dampers open. On a decrease in mixed air temperature the reverse sequence occurs.

The outside air dampers should full open before the cooling coil control valve begins to open.

If there is a separate minimum outdoor damper it will open to it's preset position when the supply fan starts and stays open until the supply fan stops. There can be a remote control for setting the minimum damper position.

There is usually a high limit control which position the outside air damper to it's minimum position if the outdoor air temperature increases to above the building temperature.

With more sophisticated systems enthalpy control is brought into play. This is when the sensible and latent heat of the outside air is measured and the mixed air dampers are positioned accordingly.

## Filter Section

The filter section exists to remove dirt and dust from the air to protect the coils and fans. The level of efficiency of filters is measured in %. The size of dirt and dust removed from the air is noted as microns.

There may be high efficiency bag filters which are often protected by lower efficiency replaceable prefilters.

The higher the efficiency of the filter the more pressure drop there will be across the filters. The more pressure drop the more fan motor HP is necesssary to develop air flow across the filters.

It is good operating procedure to measure pressure drop across clean filters and when pressure drop increases approximatey .3" WC it is time to change them.

The dirtier the filter the less supply air will be developed.

## Preheat Coil & Heating Coil

The air system may have a preheat as well as heat coil in cold climates, especially for air systems which require a higher percentage of minimum outside air. The preheat and heat coil have similar control configuration.

A controller or BAS sensing coil discharge air temperature modulates the control valve which controls glycol flow through the coil. Most coil control configuration uses a three way control valve. When flow is not desired through the coil it is diverted to the return. The supply is to the bottom of the coil and the return is from the top of the coil.

There should be a glycol circuit balance valve (CBV) in the control valve common return.

The CBV should be balanced to the maximum flow rate that the coil is designed for. The CBV may be used as an isolation valve and is equipped with a memory reset which allows the CBV to be opened only to the balance position after being closed. It is not unusual that if a CBV is located at the discharge of a circulating pump that the pump must be "OFF" before the CBV can be full closed. The CBV has 3 uses, isolating valve, flow measurement station and balancing valve.

The purpose of an hydronic balance is to ensure that coils closest to the discharge of the main glycol circulating pump do not allow too much supply glycol to bypass to the return line therefore "starving" coils in air systems further down the line.

The air system discharge air temperature is limited to a predetermined high temperature. Should the air system shut down the coil control valves will open to provide full flow through the coil.

The pump should run continuously when the outside air temperature is below a preset temperature limit.

## Preheat Coil & Heating Coil Circulating Pump

The coil circulating pump circulates glycol through the coils. The pump usually runs continuously during cold weather when the air system is "ON". This is usually determined by the BAS.

## Cooling Coil

A controller or BAS sensing coil discharge air temperature modulates the control valve which controls chilled water flow through the coil. Most control configuration uses a two way control valve.

When flow is not required it is simply stopped. If the system has two way control there must be a pressure differential bypass from supply to return usually located across the supply and return at the end of the line.

There should be a CBV installed in the return chilled water line and it serves the same purpose as explained for the preheat and heating coil.

## Direct Expansion Cooling Systems

The cooling coil outlet suction line sensor which senses refrigerant discharge temperature modulates the control valve on the inlet to the coil. This regulates the flow of refrigerant through the coil to maintain the desired discharge air temperature. On small systems this control valve is controlled open or closed.

On direct expansion refrigerant coils care must be taken in the setting of the refrigerant discharge temperature so that the refrigerant discharge air temperature is limited to a low of 14 C. If this is not so it is possible that moisture that condenses on the coil surface may freeze and eventually the coil will become blocked with ice. This can also happen if there is low air flow across the evaporator coil.

Control theory is such that a controller sensing zone temperature energizes an electric solenoid valve and a thermostatically self operating expansion valve. These are located in the inlet to the piping for the coil. The solenoid is opened on a demand for cooling and the TXV regulates the amount of liquid refrigerant that passes through the coil. The vapourized refrigerant in the tubes of the cooling (evaporator) coil absorbs heat which is transferred from the air passing over the coil tubes.

The thermostatic expansion valve is located in the refrigerant line to the cooling coil. The compressor operates while the thermostatic valve allows refrigerant into the cooling coil. When the desired coil temperature is reached the TXV throttles the flow of refrigerant to assure that only gaseous refrigerant leaves the cooling coil outlet and moves on to the suction side of the compressor. When the refigerant demand is satisfied, the solenoid valve closes. As the compressor keeps reducing the pressure on the suction line a constant pressure is reached which turns off power to the motors automatically.

On large systems it is necessary to control the compressor discharge in stages to match the loads and prevent cycling of the compressor motor. The sensor for the control of the unloading of cylinders can be based on, suction pressure of the unit, temperature of the return air or temperature of the occupied space.

## Packaged Condensing Unit

The condensing unit provides refrigerant flow to the remote evaporative direct expansion coil located in the air system. It is usually meant to be operated during summer months only but when called for, winter operating packages are available.

The unit operates automatically when:
Compressor starter switch is in the "ON" position.
Condenser starter switch is in the "ON" position.
Outdoor air temperature is above 13 C.
Air system is "ON".
Cooling coil discharge air temperature is above set point.
On a demand for cooling this sequence occurs:
Control relay energizes the solenoid valve at the DX coil.
Evaporation of refrigerant in the coil starts to raise the pressure on the suction side of the machine which energizes the control relay at the condensing unit control panel.
Control relay for the compressor and related control circuits is energized.
Control relay for the condensing unit fan control circuit is energized.

The thermostatic expansion valve is located in the refrigerant line to the cooling coil. The compressor operates while the thermostatic expansion valve allows refrigerant into the cooling coil. When the desired coil temperature is reached the TXV throttles the flow of refrigerant to assure that only gaseous refrigerant leaves the cooling coil outlet and moves on to the suction side of the compressor. When the refrigerant demand is satisfied, the solenoid valve closes and as the compressor keeps reducing the pressure on the suction line a constant pressure is reached which turns off the power to the motors automatically.

## Cooling Coil Circulating Pump

The chilled water coil circulating pump circulates chilled water through the coil. The pump usually runs continuosly during hot weather when the air system is ON. This is usually determined by the BAS.

## Humidifying Section

A humidity controller which may be located in the conditioned space or the return air duct senses the average relative hunidity. On a demand for humidity this will modulate the control valve on the steam grid or energize the wall mounted steam generator.

A high limit controller with a sensor downstream of the steam grid limits humidification to no higher than a preset value. This will prevent saturation of the supply air.

A water treatment program should be initiated to control the growth of algae in the sump tank. There should be a weekly blowdown procedure so that mineral concentration is maintained below the saturation level for the sump tank. The blowdown procedure may consist of an automatic bleed line connected to the spray header which feeds a manually adjustable stream of water to drain whenever in operation. It may also consist of an automatic sensor which monitors the mineral concentration and when it gets too high will automatically cause drainage of the sump and the addition of less concentrated raw water to the sump.

## Supply Air Fan

The supply air fan delivers air to the building and provides for temperature control and proper ventilation for carbon dioxide (CO2) control. Carbon dioxide is a result of the human breathing process. The naturally occuring level of CO2 is about 400 ppm and if it increases to 600 to 700 ppm it can make people drowsy and inattentive. The existing standard is that supply air should consist of a minimum of 20 cfm of outside air for each occupant. Due to energy conservation considerations these standards are under review.

Building supply air which consists of outside air and return air must exceed the total of building return and exhaust air. It is desireable that the building be at a neutral or slightly positive pressure relative to outside. This will stop infiltration of unconditioned unfiltered outside air into the building. This outside air will find it's way through the building envelope and doors and windows if the building is at a negative pressure relative to outside.

On air system start up the return fan should start before the supply fan and on shutdown the supply fan should stop before the return fan.

The supply air fan may be constant volume or variable air volume depending on system design.

## Wall Mounted Thermostats

Wall mounted thermostats may be pneumatic, electric or electronic. The system may have a wall mounted temperature sensor which delivers a temperature signal to a building automation system.

## Pneumatic Thermostats

Pneumatic thermostats require a 20 psi control pressure to function. This control air is supplied by a pneumatic control air compressor.

The wall mounted thermostat uses a 3 to 15 psi output signal to control either a damper, cooling coil and/or heating coil control valve to provide modulating temperature control. Most heating, ventilating and air conditioning thermostats are direct acting. This means, if the room temperature rises above the set point of the thermostat the thermostat output (psi) increases to return the room to set point temperature. If the room temperature falls below the set point of the thermostat the thermostat output (psi) decreases to return the room to set point temperature.

Usually there is a normally open heating valve which is driven closed at 8 psi. At 9 psi the heating valve is closed and the cooling function has not yet started. This is called "deadband". At 15 psi the system is calling for full cooling.

## Electric Thermostats

Electric thermostats require a low voltage to function. The thermostat is usually used for an on off function. It may position a damper, open a cooling or heating valve or energize a heating coil or DX cooling system. These thermostats are also usually direct acting.

Electric thermostats may have fan and control selector switches. If the thermostat is controlling a fan driven device a selector switch may allow the occupant to select continous or intermittent fan operation.

If continuous is selected the fan will be on and the thermostat will control heating and cooling. If intermittent is selected the fan will only energize when heating or cooling is called for.

The other selector switch may have heat, auto, cool and off positions. On heat only heat can be selected. On cool only cool can be selected. On auto, heat or cool may be selected. When thermostat temperature controls are selected off, the fan may run continuously but neither heat or cool will operate.

## Electronic Thermostats

Electronic thermostats require a low voltage to operate. The thermostat can be used for on off or modulating control functions. It may position a damper, open a cooling or heating valve or energize a an electric heating coil or DX cooling system.

These thermostats are usually direct acting.

These electronic thermostats may also send a signal to a building automation system which provides the control function.

When used as part of a building automation system it may act as a wall mounted temperature sensor with no local selection of control signal.

If provided with selector switches they would operate the same as for electric thermostats.

It may just act as sensor which deliver a signal to the BAS indicating temperature.

## Pneumatic Control Air Compressor

The pneumatic control air compressor provides the required 20 psi signal for the building temperature controls.

The system will consist of a compressor, filter station, refrigerated after dryer and pressure reducing station.

The compressor provides high pressure air to the receiver tank. The receiver tank is equipped with pressure sensing controls which control the compressor operation.

The filter station removes oil from the compressed air which is supplied to the building controls. Oil in the compressed air will cause problems for the pneumatic controls.

The refrigerated after dryer removes moisture from the compressed air. Moisture in the compressed air will cause problems with the pneumatic controls.

The pressure reducing station allows the high pressure compressed air to be reduced to that level as required by the building pneumatic controls. This pressure should be a little higher than required to compensate for losses in the system.

# COOLING SYSTEM

The mechanical equipment that may be included in a cooling water system will consist of:

Chillers
Chilled water circuit pumps
Chilled water expansion tank
Chilled water makeup
Chilled water chemical treatment
Chilled water excess pressure control loop
Condenser water circuit pumps
Condenser water makeup
Condenser water chemical treatment
Cooling tower
Air cooled condenser

## Chillers

The chillers may be reciprocating or centrifugal and the purpose is to provide chilled water for air systems cooling coils. The chilled water exists to provide a medium which will remove heat from the building air to maintain comfortable ambient conditions. The chilled water is usually supplied at 42 F but should be matched with the design requirements of the air systems cooling coils. Prior to the cooling season the chiller should be thoroughly inspected before startup. The inspection must be done by service personnel qualified in start up procedures for this equipment.

## Theory of Operation

In the refrigerating cycle, heat from air passing over the cooling coils raises the water temperature which is circulated through the evaporator. The heat of such passes through the chiller coils, raises the temperature of the liquid refrigerant to it's boiling point and evaporates it into a gas.

Note: One ton of refrigeration is the rate of heat removal which will produce one ton of ice at 32 F from water at 32 F in 24 hours.

The three main parts of a chiller are:

**Evaporator** section which removes heat from air passing over the air system cooling coils.
The evaporator consists of a series of coils in which the liquid is evaporated into a gas by absorbing the return chilled water

heat at a low temperature after it's pressure has been reduced by passing through the expansion coil.

**Compressor** section maintains a lower pressure in the evaporator than in the condenser so that the refrigerant liquid may expand and drop in pressure while passing through the small opening of the expansion valve. This raises the pressure of the evaporated liquid to a point where the condensing temperature is above the temperature of the cooling water available.

Note: The refrigerant serves as a heat transferring medium with the heat finally disposed of by the condenser cooling water.

The compressor draws gas from the evaporator at suction pressures of from 10 to 30 psi and compresses it to between 150 to 170 psi before delivering it to the condenser. When the gas from the evaporator coil enters the suction of the compressor, it is at, or near the saturation temperature of the condenser cooling water. The process of compression raises it's temperature above the cooling water temperature and thus allows heat to flow from the refrigerant gas to the condensor cooling water. The abstraction of heat from the refrigerant gas brings it's temperature below the boiling point corresponding to the condensor pressure and it condenses back to the liquid form ready to again pass through the expansion valve and continue the operation.

**Condenser** section cools the hot gas received from the compressor to a point where it will condense into a liquid. The condenser may be air cooled or water cooled.

The air cooled condenser simply has air blown through rows of tubes containing the gas from the compressor.

The water cooled condenser complete with cooling tower has condenser water circulating in small inner pipes and the gas and liquid refrigerant in the space between the inner and outer pipes. The hot gas and the water travel in an opposite direction so that the coldest refrigerant gas leaves the condenser at the point where the condenser water enters. The heat in the condenser water is then removed as it flows through the cooling tower.

## Chilled Water Circuit Pumps

The purpose of the chilled water pumps is to circulate chilled water from the chiller to the air systems cooling coils and back to the chiller. Pump flow is designed to match the required flow rate of both the air system cooling coils and the chiller. Pumps are now usually installed with circuit balance valves (CBVs) at the discharge of the pump. The CBV allows water flow to be measured and balanced to design flow rates. The pumps must be in operation for the chiller to start.

## Chilled Water Expansion Tank

The chilled water expansion tank allows for fluctuation in the volume of water in the system as the water temperature changes.

Under normal operating conditions the tank level should not rise above the top or drop below the bottom of the sight glass. The sight glass is mounted on the side of the expansion tank.

A compressed air line connection is provided to the tank which allows for pressurization of the expansion tank.

The pressurization of the system serves three purposes.

It maintains the chilled water circuit at the design pressure.

It provides a positive head on the suction side of the circulating pumps ensuring air is not drawn into the system.

It saves wear and tear on the pumps on startup.

## Chilled Water Makeup

The makeup water system consists of a bypass valve, shut off valves and a pressure regulating valve (PRV). One side of the system is piped to the domestic cold water system and the other side is piped to the chilled water expansion tank at the suction side of the circulating pumps.

If the PRV malfunctions and allows more makeup water into the system than necessary, then the pressure relief valve will open and allow the system to drain until the excess pressure is released.

The relief valves are each set above the system design operating pressure and should not be adjusted.

The shut off valves are used to isolate the PRV should it ever malfunction. The bypass valve may be opened manually to add water whenever the expansion level drops too low or until the PRV is repaired and put back into service.

Double check valves are provided at the domestic cold water line connection to the pressure regulating valve to prevent chemically treated water from entering the cold water system, in the case of reverse flow.

## Chilled Water Chemical Treatment

Chemicals necessary to treat the chilled water may be added manually or automatically. It is usually a manual function in most applications. However, on more sophisticated systems water may be automatically tested and chemicals added automatically.

They may be added to either the chilled water supply or return lines.

It is suggested that chemical tests be performed on a weekly basis to determine chemical concentration and that conditions be maintained as directed by a chemical treatment specialist. Chemical treatment is necessary to inhibit rust and corrosion of the water side of all chilled water components.

## Chilled Water Excess Pressure Control Loop

A bypass automatic control which senses the differential pressure between the supply and return line is provided to maintain a constant pump head pressure at the circulating pumps. The control positions a normally closed two port valve towards the open or closed position. It responds to increases or decreases in differential pressure in the cooling system. The increased pressures developed at low cooling demands must be controlled. The bypass control valve is located in a crossover pipe, between the supply and return water lines. This crossover may be located at the pumps or at the end of the supply and return line.

## Condenser Water Circuit Pumps

The purpose of the condenser water circuit pumps is to circulate condenser water from the chiller to the cooling tower and from the tower sump back to the chiller.

The system may be designed with two pumps. Both pumps are usually of the same size and one or both pumps may operate. The pumps must be operating before the chiller will run.

## Condenser Water Makeup

The automatic makeup for the condenser water system consists of float valve in the cooling tower sump. When the sump water level decreases the float valve lowers, which in turn opens the makeup water valve. Domestic cold water is introduced into the sump to increase the level back to normal. During tower operation much water is evaporated resulting in a good deal of makeup water being introduced.

## Condenser Water Chemical Treatment

The condenser water chemical treatment may consist of a chemical pot feeder located on the discharge side of the condenser water circulating pumps. Holding tanks contain the chemical which is automatically fed into the system in a metered fashion by a chemical feeder pump.

It may also be a blowdown system. If manual or automatic tests register a chemical imbalance the system may simply blow down and be replaced by domestic water.

It may also be and automatic blowdown system.

## Cooling Tower

The cooling tower serves to remove heat from the condenser water. This is done in a temperature controlled manner.

The cooling tower fans may run continuously or may be automatically energized to maintain the condenser water at design temperature to suit the chillers.

## Air Cooled Condenser

The air cooled condenser removes heat from the refrigerant unit discharge gases allowing the refrigerant gases to be condensed and recirculated. The condenser is roof mounted and has air blown through rows of tubes containing the gas from the compressor.

All controls are an integral part of the air cooled condenser packaged unit. The condenser fans are cycled on and off automatically to remove the heat from the refrigerant compressor discharge gases.

# HEATING SYSTEMS

The mechanical equipment included in a heating water system may consist of:

Hot water boilers
Steam boilers
Primary and secondary circuit heating water pumps
Primary and secondary circuit heating water expansion tanks
Primary and secondary circuit makeup water system
Primary and secondary circuit heating water chemical treatment
Heat exchangers
Steam boiler feedwater chemical treatment
Radiation heating water temperature control loop
Heating water excess pressure control loop
Ceiling hung and cabinet unit heaters
Fan coil units
Ducted force flow units
Convector radiation heating water system
Glycol circuit circulating pumps

Glycol circuit charging tanks
Glycol circuit expansion tanks

## Hot Water Boilers

The purpose of the boilers are to provide heating water for the various building mechanical heating components. The heating water allows heat to be transferred from the primary fuel, gas, oil, coal and electricity by means of the hot water to the zone to maintain comfortable ambient conditions.

The heating components supplied by the boiler heating water are heat exchangers, ceiling hung unit heaters, cabinet hot water unit heaters, fan coil units, ducted force flow units, convector radiation system and air system heating coils.

The heating water may be supplied at a constant or controlled varying temperature.

## Steam Boilers

The purpose of steam boilers are to provide steam for the various building mechanical heating components and humidification requirements.

The steam allows heat to be transferred from the primary fuel, gas, oil, coal or electricity by means of the steam to maintain comfortable ambient conditions.

The mechanical components supplied by the boiler steam are, heat exchangers, convector radiation system, ceiling hung unit heaters, cabinet unit heaters, air system heating coils and air system humidifiers.

The steam is supplied at a constant pressure.

## Hot Water & Steam Boiler Terminology

There are common terms used in the language of hot water boilers and it is suggested operators be familiar with them.

Low pressure boiler - operates at pressures not above 15 psi.

High pressure boiler - operates at pressures above 15 psi.

Water space - the parts of a boiler that contain the water.

Steam space - the space above the water level in a steam boiler.

Heating surface - the heating surface of a boiler consists of all parts of a boiler through which heat is transferred from the burning fuel to the water. It includes all parts of the boiler plates and tubes which have water on one side and which are swept by fire or hot gases on the other.

Combustion chamber - the part of the boiler furnace where fuel gases are mixed with air and burned.

Water line - the level of water in the boiler.

Combustion gases - the hot gaseous products from the fire. These are referred to as flue gas.

Stack - the stack or chimney is the vertical passageway which discharges the flue gases to the atmosphere some distance above the boiler.

Draft fans - these are mechanically driven fans used to either blow air into the furnace via the windbox or to draw the combustion gases from the boiler and discharge them to the stack.

Fittings - these are devices fitted to the boiler which makes it easier and safer to operate.

Gauge glass - this fitting, which consists essentially of a strong glass tube is used to show the level of water in the boiler.

Water column - this is a chamber which is connected to the boiler and to which is attached the gauge glass.

Safety valve - this fitting prevents the pressure within the boiler from exceeding safe limits. It will open and release water or steam when the pressure reaches set point.

Steam gauge - this fitting indicates the pressure of the steam within the boiler.

Check valve - this is a valve which allows flow in one direction only. It will close automatically in the event of reverse flow.

Blowoff valves - these valves can be opened while the boiler is in operation in order to blow out sludge and sediment from the boiler. The boiler may also be drained if necessary by means of these valves.

Low water cutoff - this is a safety device which will cut off the boiler burner if the water level in the boiler becomes dangerously low.

Feed water - this is water that is fed to the boiler to replace water or steam drawn or leaked from the boiler.

Condensate - when the steam from a boiler is used in a heating system, process, engine or turbine, it will give up it's heat and condense to water which can be returned to the boiler. This water is known as condensate.

## Primary and Secondary Circuit Heating Water Pumps

The purpose of the primary and secondary circuit heating water pumps is to provide heating water to heat exchangers, ceiling hung unit heaters, cabinet hot water unit heaters, fan coil units, ducted force flow units, convector radiation system and air system heating coils. Pump flow is designed to match the required flow rate of the heating components and the boilers. Pumps are now usually installed with circuit balance valves (CBVs) at the discharge of the pump. The CBV allows water flow to be measured and balanced to design flow rates. The pumps must be in operation for the boilers to fire.

## Primary and Secondary Circuit Heating Water Expansion Tanks

The primary and secondary circuit heating water expansion tanks allow for fluctuation in the volume of water in the system as the water temperature changes.

Under normal operating conditions the tank level should not rise above the top or drop below the bottom of the sight glass. The sight glass is mounted on the side of the expansion tank.

A compressed air line connection is provided to the tank which allows for pressurization of the expansion tank.

The pressurization of the system serves three purposes.

It maintains the heating water circuit at the design pressure.

It provides a positive head on the suction side of the circulating pumps ensuring air is not drawn into the system.

It saves wear and tear on the pumps startup.

## Primary and Secondary Circuit Makeup Water System

The primary and secondary circuit makeup water system consists of a bypass valve, shut off valves and a pressure regulating valve. One side of the system is piped to the domestic cold water system and the other is piped to the heating water expansion tank at the suction side of the circulating pumps.

If the PRV malfunctions and allows more makeup water into the system than necessary, then the pressure relief valve will open and allow the system to drain until the excess is released.

The relief valves are each set above the system design operating pressure and should not be adjusted.

The shut off valves are used to isolate the PRV should it ever malfunction. The bypass valve may be opened manually to add water whenever the level drops too low or until the PRV is repaired and put back into service.

Double check valves are provided at the domestic cold water line connection to the pressure regulating valve to prevent chemically treated water from entering the cold water system, in the case of reverse flow.

## Primary and Secondary Circuit Heating Water Chemical Treatment

Chemicals necessary to treat the heating water may be added manually or automatically. It is usually a manual function in most applications. However, on more sophisticated systems water may be automatically tested and chemical added automatically.

They may be added to either the heating water supply or return line.

It is suggested that chemical tests be performed on a weekly basis to determine chemical concentration and that conditions be maintained as directed by a chemical treatment specialist. Chemical treatment is necessary to inhibit rust and corrosion of the water side of all heating water components.

## Heat Exchangers

The heat exchanger provides either heated water or glycol for various building mechanical heating components. The exchanger may be a water to water, water to glycol, steam to water or steam to glycol.

Heating components supplied by the heat exchanger are ceiling hung unit heaters, cabinet hot water unit heaters, fan coil units, ducted force flow units, convector radiation system and air system heating coils.

## Steam Boiler Feedwater Chemical Treatment

Liquid in a steam system can alternate between the steam side of a system and the atmospheric condensate side resulting in oxygen corrosion. The steam may be used for humidification purposes and quantities of raw water must be added.

It is essential that feedwater chemical tests be done on a daily basis. Specialists in this field should analyze the chemical treatment requirements and provide detailed instructions on the use of their particular products and on the procedures for testing.

The chemical supplier should provide a test kit and log sheets to record the results of these tests and corrective action to be taken.

Generally the PH of the boiler water should be alkaline with a reading around 8.5 to 10. Acidic conditions will result in corrosion of the boiler tubes. Whenever the PH gets too low the addition of sodium hydroxide is usually made to correct the condition. Water softeners are used to remove calcium carbonates from the feedwater as these will result in severe scaling of the boiler internals.

## Radiation Heating Water Temperature Control Loop

A control sensing supply water and outside air temperature is provided to vary the heating water temperature supplied to the system. The control will modulate a mixing valve which mixes supply and return water to achieve the necessary temperature.

As the outdoor air temperature increases the heating water supply temperature decreases. As the outdoor air temperature decreases the heating water supply temperature increases. A common reset schedule is at -30 C outside air the supply water temperature is 82 C and at 15 C outside air temperature the supply water temperature is 30 C.

## Heating Water Excess Pressure Control Loop

A bypass automatic control which senses the differential pressure between the supply and return line may be provided to maintain a constant pumphead pressure at the circulating pumps. The control positions a normally closed two port pneumatic valve to the open or closed position. It answers to increases or decreases in differential pressure in the heating system. The fluctuations in system differential pressure are a result of the opening and closing of two way control valves throughout the building. The increased pressures developed at low heating demands must be throttled as control valves open and close causing objectionable noise. The bypass control valve would be located in the crossover pipe between the supply and return water lines on the discharge side of the circulating pumps and the boiler return.

## Ceiling Hung and Cabinet Unit Heaters

The purpose of the ceiling hung heaters is to provide heating for the immediate area in which it is located. The heater uses 100% return air for ventilating purposes.

The ceiling hung unit may consist of an electric or pneumatic wall mounted thermostat, a hot water, electric, glycol, steam heating coil or gas fired burner section. The is also a supply air fan which moves air over the heating section and automated controls.

## Fan Coil Units

The purpose of the fan coil unit is to provide heating and cooling for the immediate area in which it is located.

The fan coil unit uses 100% return air for ventilating purposes.

The fan coil may be comprised of a wall mounted electric or pneumatic thermostat, a hot water, electric, glycol or steam coil, chilled water cooling coil, supply fan and automatic controls.

## Ducted Force Flow Units

The purpose of the ducted force flow unit is to provide heating and or cooling for the area in which it's supply air diffusers serve.

The ducted force flow unit can use 100% return air or a mix of return and outside air for ventilating purposes.

The force flow unit is comprised of a wall mounted thermostat either pneumatic or electric, hot water heating coil, chilled water cooling coil, supply air fan and automatic controls.

## Convector Radiation Heating Water System

The convector radiation heating water system provides heating for the perimeter of the building. The finned convectors are located throughout the building, wall mounted on the perimeter of the exterior rooms.

The convector radiation system may consist of a wall mounted thermostat either electric or pneumatic or convector mounted self contained thermostatic valves, finned convectors and control valves either modulating pneumatic or electric two position.

## Glycol Circuit Circulating Pumps

The purpose of the glycol circuit pumps is to provide a mixture of glycol and hot water to heating components throughout the building. The pumps move glycol through the secondary of a heat exchanger or through the boilers.

## Glycol Circuit Charging Tank

The glycol circuit charging tank provides a means to add glycol to the system when necessary. Glycol may be stored in a tank under a higher pressure than the operating pressure of the glycol heating system. It is pumped into the glycol heating system when necessary to add glycol to the system. The glycol system will be a mixture of glycol and water.

The automatic glycol makeup system consists of a makeup valve, shut off valves and a pressure reducing valve. One side of the system is connected to the pressurized glycol charging tank and the other is connected to the line from the glycol heating system to the expansion tank. The pressure reducing valve maintains a constant pressure on the heating system to the expansion tank. The pressure reducing valve maintains a constant pressure on the heating system by allowing glycol to enter the system when system pressure drops below the pressure reducing valve setting.

The shutoff valves are used to isolate the PRV should it malfunction. The bypass valve is opened manually to add glycol solution whenever the expansion tank level drops too low.

## Glycol Circuit Expansion Tank

The glycol circuit expansion tank allows for fluctuation in the volume of glycol as the temperature changes.

The expansion tank size is as designed to allow for full operating temperature ranges in the glycol. Under normal operating conditions the tank level should not rise above the top or drop below the bottom of the sight glass. The sight glass is mounted on the side of the expansion tank.

This pressurization serves three purposes:

It maintains the glycol circuit at the design pressure.

It provides a positive head on the suction side of the circulating pump ensuring air is not drawn in through the automatic vents.

It saves wear and tear on the pump during startup.

# PLUMBING

The building plumbing system may consist of:

Domestic water
Booster pumps
Pressure reducing station
Non freeze hose bibbs
Domestic hot water
Sanitary drainage
Storm drainage
Sumps
Gas service

## Domestic Water

A domestic cold water main will be provided to the building. The main is connected to a water station consisting of a water meter and shut off valves. The main building shut off valve is located directly in front of the water meter. A bypass valve may be provided with a sealed valve only to be used with the water authorities permission.

The domestic cold water service is provided for the purpose of:

Cold water supply to sinks, lavatories, water closets, baths and showers.

Boiler makeup water.

Domestic hot water makeup.

Condenser water makeup.

Fire protection system.

## Booster Pumps

The pumps boost the city water pressure to that necessary to provide adequate water flow in all parts of the building.

## Pressure Reducing Station

When the the booster pumps raise the system pressure to a level to satisfy the system at the highest point the pressure may then be too high for the lower points. The pressure reducing stations using a pressure regulating valve with isolation valves allows for automatic reduction of pressure for each floor.

## Nonfreeze Hose Bibbs

Nonfreeze hose bibbs may be provided as exterior water tap connections at various locations outside the building. These water valves are designed to shut off the flow of water inside the building before the line passes through the outer wall preventing winter freeze up of the water line.

## Domestic Hot Water

The domestic hot water system heats cold water to produce domestic hot water for building use. It is suggested taht the domestic hot water be maintained at a minimum 38 C to satisfy health code requirements. Ensure the temperature of the domestic hot water is not controlled too high as scalding could occur.

The domestic hot water supply system may consist of:

Gas fired domestic boilers.
Electric boilers.
Gas fired hot water heaters.
Electric hot water heaters.
Recirculating pump.
Heat exchanger (water to water)
(steam to water)
(glycol to water)
Storage tank.
Makeup water system.

## Sanitary Drainage

The sanitary drain provides a drain for:

Water closets.
Urinals.
Baths.
Showers.
Lavatories.
Sinks.
Floor drains.
Clothes washers.
Dishwashers.
Drinking fountain overflow.
Sumps.

These components discharge into the main sanitary drainage line which carries the sewage out of the building. The main building cleanout will be located before the sewer line passses through the building outer foundation wall. Access to this cleanout can be made from the floor metal cover.

The sewer line also has vent lines attached to it which vent directly up through the roof. These vents permit excess sewer gas to be released to atmosphere amd prevent siphonage of fixture trap seals.

## Storm Drain

The storm drain consists of roof drains connected to rain water leaders which are drained to the storm drainage main. Also surface water from around the outside perimeter of the building seeps into the underground weeping tile collection system which then drains into the storm drain via a sump or trap which is complete with a backwater valve. Surface water from car park areas may also drain into catch basins which are connected directly to the storm sewer.

It may also be that the accumulated water drains back into the building and collects in the sump. In older systems the waste water may be drained to the sanitary sewer system.

The main building cleanout will be located before the storm line passes through the building outer foundation wall. Acess to the cleanout can be made from a metal plate cover on the floor.

## Sump

Waste water collected from the sanitary sewer system, mechanical room floor drains, weeping tile drainage system empty into a sump collection pit.

From the collection pit the sewage is held until the level reaches a high enough point at which time it will flow into the outlet line due to the flow of gravity.

It may be that from the collection pit pumps remove the waste water into the sanitary sewer system.

There are usually two pumps. The first pump is activated when the collection pit level rises. If the first pump cannot maintain the water level the second pump will energize.

## Gas Service

A natural gas service line may be provided to the building.

A building main shutoff valve is located directly before the gas meter. The gas service is provided for all gas fired equipment located in the building.

A shutoff valve, pressure regulating valve and pilot valve are provided at each gas fired unit.

The gas may be supplied from local utilities or propane tanks at low, medium or high pressure.

Pressure regulating valves are installed on the gas fired appliances to reduce the gas pressure to the usual burner operating pressure of 3" WC.

# FIRE PROTECTION

The fire protection system for the building may be comprised of both manual and automatic systems which are:
Hand fire extinguishers

Automatic sprinkler systems
Wet pipe
Dry pipe
Preaction
Deluge
Combined dry pipe and preaction
Antifreeze type
Central carbon dioxide
Dry chemical

Water distribution facilities
Wet standpipe with siamese connections
Fire hose cabinets
Fire pumps
Alarms

## Hand Fire Extinguishers

Hand fire extinguishers are provided to handle fires during early stages where the hazard requires specific types of extinguishers. The building operator should be familiar with the locations of all fire extinguishers, their type and what class of fire they are used for.

The extinguishers provided are:

Carbon dioxide which are intended to be used on class B and C type fires involving flammable liquids, gasoline, greases and electric powered equipment and switchgear.

Multipurpose dry chemical which are intended to be used on class A, B and C type fires involving wood, paper, flammable liquids, gasoline, grease and electric powered equipment and switchgear.

Water pump which are intended to be used on class A type fires involving paper, wood or cardboard.

## Wet Pipe Sprinkler System

This system uses automatic sprinklers attached to a piping system which contains water and is connected to a water supply enabling the water to discharge immediately from sprinkler heads released by a fire. The location of the main sprinkler valve should be known by the operator.

The system may be at city water pressure or maintained about 25 psi above city water pressure.

This feature has ben provided to reduce the frequency of nuisance false alarms when there is a surge of pressure in the city lines. To provide over pressure there is a small high presure pump connected from the city line side across the sprinkler valve to the sprinkler distribution system. A pair of pressure gauges are provided to show the pressure on each side of the sprinkler valve. To develop the pressure requires manual operation. The isolation valves for the pump must be opened and the pump turned on. Allow the pump to run until there is a 25 psi differential pressure across the gauges. While this may take several minutes do not let the pump run unattended to ensure there is no chance of accidental damage to the sprinkler system. When adequate over pressure is obtained shut off the pump and close the isolation valves.

Once a sprinkler system has been activated and no further fire hazard exists it is necessary to close the manual gate valve ahead of the sprinkler valve.

The sprinkler heads that were released should be replaced and the main valve opened to make the sprinkler system operational again.

## Dry Type Sprinkler System

The system uses automatic sprinklers attached to a piping system which may contain air or nitrogen under pressure. The escape of the air or nitrogen from the release of a sprinkler permits the water pressure to open a valve known as a dry pipe valve. The water then flows into the pipng system and out the opened sprinklers. The building operator should know where the main sprinkler valve is located.

## Preaction Sprinkler System

The preaction system employs automatic sprinklers attached to a piping system containing low pressure air and uses a supplemental fire detection system. Actuation of the detection system opens a valve which permits water to flow into the sprinkler piping system and discharge through any sprinkler heads which may have opened.

The low air pressure in the piping system serves to activate a low pressure alarm if vandals should release any sprinkler head with a cigarette lighter.

In the event the system has been activated, the piping must all be drained and new sprinkler heads installed as required.

## Deluge Sprinkler System

The deluge system employs open sprinklers attached to a piping system connected to a water supply through a valve which is opened by a supplemental fire detection system. It is located in the same area as the sprinklers. When this valve opens water flows into the piping system and discharges from all the sprinklers. Complete water spray coverage is the result. The operator should know where the automatic sprinkler valve is located.

In the event that the supplemental fire detection senses a fire condition, an alarm will sound but opening of the sprinkler valve will be delayed for a preset time. This will allow the operator time to quickly assess the situation and to abort the tripping of the sprinkler valve which would otherwise dump water in the area possibly causing unnecessary water damage.

## Combined Dry Pipe and Preaction Sprinkler System

In the event of a fire alarm from the supplemental detectors in the area an alarm condition is indicated on the control panel.

Opening of the sprinkler valve will be delayed for a preset period of time. This gives the operator time to quickly assess the alarm condition and abort the opening of the sprinkler valve if it happens to be a false alarm. If no action is taken during the delay the sprinkler valve opens along with the special air vent valves and the system goes into action, spraying water out of any heads that have released.

## Antifreeze Sprinkler System

The system uses automatic sprinklers which are attached to a piping system containing an antifreeze solution and connected to a water supply. The antifreeze solution followed by water, discharges immediately from sprinklers opened by a fire.

## Carbon Dioxide System

This system is provided for specialty areas which are especially hazardous. A bank of carbon dioxide tanks will be located adjacent to the area.

These tanks are manifolded in such a manner that predetermined quantities of $CO_2$ are available for each especially hazardous room. An auxiliary fire sensing system in each area will sound an alarm in that area to warn anyone in the space that the $CO_2$ discharge system is to be activated.

This is a very toxic gas which can cause brain damage or death if significant quantities are inhaled by a person in the room when the discharge occurs. Once initiated at the manifold the $CO_2$ is fed to all nozzles on that system until the tanks are empty.

There is usually an abort switch located just outside the door to the room for the operator to prevent the release of $CO_2$ in case of false alarm.

A special tank weighing station may be provided to periodically check each tank so that none are emptied because of a slow leak.

## Dry Chemical System

This system is provided for kitchen exhaust areas and consists of a tank of dry chemical which is piped to a series of nozzles over the most hazardous areas in the food preparation area.

Special sensors located under the hood trigger the release of the dry chemical from the tank.

At the same time a solenoid valve on the gas line serving the cooking equipment shuts off.

## Water Distribution Facilities

These facilities are provided to allow use of a large quantity of water to fight fires which are in advanced stages. The following piping and equipment is installed.

## Wet Standpipe with Siamese Connections

There may be one or more standpipes which are connected to a siamese connection at ground level on the exterior of the building.

A check valve between the siamese fitting and the standpipe with it's normal source of water makes it possible for that portion of the piping exposed to possible freezing conditions to be dry. A small hole in a low point on this outside line assures that no water can collect in this dry part of the line.

At each floor level there is a valve with outlet having threads compatible with the hoses of the local fire department. These valves may be in the stairwells, in special cabinets or in conjunction with equipment in the fire hose cabinet.

There is often a fire pump to boost the city water pressure enough to meet fire code requirements.

The siamese connections at ground level are available to the fire department pumper truck to connect to. The firemen can then connect their hoses at any of the fittings provided on the riser to battle fire at any floor.

## Fire Hose Cabinets

The cabinets are provided to contain hose racks, hose valves, nozzles, hose piping and valves. The shutoff valve must be opened when the hose is to be used. The building operator should become familiar with the locations of all cabinets.

## Fire Pumps

The pumps are supplied complete with cabinet, motor starter, circuit breaker, disconnect switch and other control devices. The fire pumps activate automatically only under a fire condition. It should be used as a pressure maintenance pump.

The separate jockey pump as described in the wet pipe system should be used to maintain the desired fire protection system pressure. When a fire hose cabinet valve is opened, the riser pressure drops which activates the fire pumps.

## Alarms

A water activated gong located on the side of the building and an electrical alarm will sound when there is water flow through the sprinkler system.

# TROUBLE EVALUATION PROCEDURES

## Hot Water and Steam Boilers

1 Is the expansion tank 3/4 full of water?

2 Are the heating water circulating pumps on?

3 Are the makeup water system isolating valves open?

4 Is the gas supply shut off valve open?

5 Is the control panel energized?

6 Is the pilot light firing?

7 Is the automatic feedwater chemical injection system functioning?

8 Is the condensate receiver tank float switch level control functioning?

## Heat Exchangers

1 Are the primary and secondary side isolating valves open?

2 Are the primary and secondary circulating pumps on?

## Expansion Tanks

1 Is the tank charged?

2 Has the air been bled from manual air vents?

3 If the tank overfills reasons could be
   - pressure regulating valve leaks.
   - pressure regulating bypass valve leaks.
   - there is an air leak on the expansion tank.

4 Is the makeup water system functional?

## Radiation Heating Water Temperature Control Loop

1. Is there power or control air to the reset controller?
2. Is the mixing valve sticking?
3. Are the mixing valve isolating valves open?

## Ceiling Hung and Cabinet Unit Heaters Fan Coil Units and Ducted Force Flow Units

1. Is the panel circuit breaker on?
2. Is the local disconnect on?
3. Is the thermostat set to the desired set point?
4. Are the heating coil isolating valves open?
5. Are the cooling coil isolating valves open?
6. Are the fan drive belts at correct adjustment?

## Convector Radiation Heating Water System

1 Is the thermostat set to the desired set point?

2 Are the isolating valves for each convector section open?

3 Are the heating water pumps running?

4 Is the air bled off at the highest point of each piping run?

## Chilled Water Systems

1 Are the chilled water pumps on?

2 Are the chilled water pumps isolating valves open?

3 Are the condenser water pumps on?

4 Are the condenser water pumps isolating valves open?

5 Is the cooling tower on?

6 Is the air cooled condenser on?

7 Is the outside air temperature above the preset low limit?

8 Is the expansion tank charged?

9 Is the makeup water system functional?

10 Is the pressure equalization valve operational?

## Air Systems

1 Are both the supply and return fan energized?

2 Is the mixed air section operating at 15 C?

3 Is the mixed air section neutral or slightly negative as required?

4 Is the minimum outside air damper opening?

5 Is the cooling coil functional?

6 Is the heating coil functional?

7 If the system is a variable air volume type is the return fan tracking the supply fan?

8 Have VAV box minimum set points been set at 25%?

9 Is the control air compressor functional?

# MAINTENANCE SUGGESTIONS

These maintenance suggestions can, for the most part, be done by the building operating and maintenance staff. Some maintenance procedures will have to be undertaken by skilled tradesmen, but it is left to the discretion of the building superintendent as to when they should be called in. The suggested frequency of tasks is as noted.

Legend

| | |
|---|---|
| Daily | D |
| Weekly | W |
| Monthly | M |
| According to manufacturer | AM |
| When necessary | WN |

## Air Compressors

| | | |
|---|---|---|
| 1 | Change crankcase oil. | AM |
| 2 | Clean fins on heads, aftercooler, cylinder. | M |
| 3 | Clean intake air filter. | W |
| 4 | Check drive belt. | D |
| 5 | Check cooling water flow. | D |
| 6 | Check oil level. | D |
| 7 | Blowdown pressure tank. | D |

## Air Cooled Condensing Units

| | | |
|---|---|---|
| 1 | Lubricate motor bearings. | AM |
| 2 | Clean fan blades. | M |
| 3 | Clean coil fins. | M |
| 4 | Check for rotation and tip clearance. | W |
| 5 | Check condenser air intake. | D |
| 6 | Check for refrigerant leaks. | D |
| 7 | Check for noise or vibration. | D |

| | | |
|---|---|---|
| 8 | Check drive coupling alignment. | D |
| 9 | Check fan drive belts. | D |

## Boilers

| | | |
|---|---|---|
| 1 | Check pressure relief valve. | D |
| 2 | Check controls. | D |
| 3 | Clean water strainers. | WN |
| 4 | Clean gauge glass. | WN |
| 5 | Check low water cutoff. | D |
| 6 | Blowdown boiler. | D |
| 7 | Visually check for cracks or leaks. | D |
| 8 | Check burner nozzle and flame colour. | D |
| 9 | Check water level. | D |
| 10 | Check control linkages. | D |
| 11 | Check electrodes. | D |
| 12 | Check float level mechanism. | D |
| 13 | Test and record water treatment. | D |

14 Check and record boiler operating temperature and pressures. D

## Chillers

1 Record operating pressures and temperatures. D

2 Check oil reservoir level on sight glass. D

3 Check oil reservoir temperature. D

4 Check purge sight glass for water. D

5 Check for oil leaks. W

6 Check operation of low oil pressure cutout. W

7 Check operation of condenser high pressure cutout. W

8 Check operation of vanes. W

9 Check operation of chilled water low temperature cutout. W

10 Check operation of refrigerant low temperature cutout. W

11 Inspect purge system. W

12 Inspect refrigerant float chamber. W

13 Inspect control centre. W

14 Inspect starting equipment. W

15 Check operation of low oil temperature switch. W

16 Check operation of cooler low pressure cutout. M

17 Check and record frequency of purge pump operation. M

18 Change oil and oil filter. AM

19 Change refrigerant filter. AM

20 Change volute drain filter. AM

21 Check compressor bearings. AM

22 Inspect and clean cooler tubes. AM

23 Check for refrigerant leaks. AM

24 Inspect and clean condenser tubes. AM

25 Change compressor oil and filter. AM

26 Change filter drier in oil return system. AM

27 Check the purge drive belt and lubricate purge motor. AM

28 Clean purge drum. AM

## Cooling Towers

| | | |
|---|---|---|
| 1 | Change reservoir oil. | AM |
| 2 | Check and clean fan blades. | WN |
| 3 | Drain reservoir pan. | M |
| 4 | Flush the pump and discharge oil. | M |
| 5 | Clean sump and header strainers. | M |
| 6 | Check oil reservoir for water and dirt. | M |
| 7 | Check and clean spray nozzles. | D |
| 8 | Check fan rotation and tip clearance. | D |
| 9 | Check bleedline. | D |
| 10 | Check water treatment levels. | D |
| 11 | Check makeup water float valve. | D |
| 12 | Check water level in sump. | D |
| 13 | Check oil level. | D |
| 14 | Check oil seals. | D |
| 15 | Check for noise and vibration. | D |

## Pneumatic and Electric Controls

| | Task | Frequency |
|---|---|---|
| 1 | Check safety controls and ensure shut down is activated. | D |
| 2 | Check transmitter calibration. | M |
| 3 | Check humidity controller calibration and clean humidity element. | M |
| 4 | Check that backdraft damper operates without binding. | W |
| 5 | Ensure spring tension is adequate to close dampers. | W |
| 6 | Clean contacts and lubricate bearings on valve actuators. | M |
| 7 | Check and clean slidewire on electric thermostats. | M |
| 8 | Clean and lubricate pneumatic damper operators. | M |
| 9 | Open nozzle, flapper and restrictor on pneumatic thermostats. | M |
| 10 | Check damper arm travel and close off. | M |
| 11 | Lubricate damper bearings. | M |
| 12 | Clean all points and contacts on control system. | M |
| 13 | Clean and set pressure controls. | M |
| 14 | Check control valves for positive shutoff. | M |

15 Clean and lubricate valve stems. WN

16 Adjust and replace packing in control valves. WN

17 Replace air pressure regulator filter. AM

18 On refrigerated after cooler, clean condenser coil. AM

19 On refrigerated after cooler, lubricate motor. AM

## Cooling and Heating Coils

1 Inspect coil fins for build up. M

2 Clean coil fins. WN

3 Check for corrosion. M

4 Ensure drain pan and line are clean. D

## Domestic Hot Water Heater

1 Check for leaks. D

2 Check that barometric flue damper is functional. D

3 Check electrical connections. D

4 Test low water cutoff. W

| | | |
|---|---|---|
| 5 | Check aquastat setpoint and water temperature. | D |
| 6 | Check pressure relief valve. | W |

## Emergency Generator

| | | |
|---|---|---|
| 1 | Check crankcase oil and filter. | M |
| 2 | Check governor. | M |
| 3 | Check spark plugs. | M |
| 4 | Lubricate. | AM |
| 5 | Check air cleaner and breather. | M |
| 6 | Check auto startup. | M |
| 7 | Check and note operating conditions. | M |
| 8 | Check electrical connections. | W |
| 9 | Check fan belt. | W |
| 10 | Check radiator cooling level. | W |
| 11 | Check and clean radiator intake screen. | W |
| 12 | Check battery cables and connections. | W |
| 13 | Check battery acid level. | W |

14 Check gas piping, manifold and flexible connectors. W

## Electric Motors

1 Record amperage and voltage. M

2 Check terminals for corrosion or loose leads. M

3 Inspect commutator and clean if necessary. M

4 Inspect brushes for wear and replace if necessary. M

5 Blow out motor windings with compressed air. M

6 Have motor windings megger tested to check insulation breakdown. AM

## Expansion Tanks

1 Check tank and system for leaks. D

2 Check expansion tank gauge level. D

3 Check makeup system operation. D

4 Check tank for corrosion. D

5 Clean gauge glass. M

## Filters

1 Check reading on filter gauge. D

2 Visually inspect filters. D

3 Clean filters, allow to dry and coat with filter oil. AM

## Finned Radiation

1 Check fins remove paper or other refuse. W

2 Clean fins. WN

3 Check for leaks around air vents, expansion joints and radiation valves. M

4 Bleed off air from each air vent. M

5 Repair bent or damaged fins. WN

## Grease Interceptors

1 Inspect depth of grease. W

2 Add chemicals to break down grease. AM

# ENERGY CONSERVATION CONSIDERATIONS

This manual deals with various measures which can be considered to save energy in a facilities mechanical and electrical systems. It will take specialized engineering help to implement many of these considerations but some can be implemented in house.

## Air Systems

1 Dual duct constant volume conversion to dual duct variable air volume operation.

2 Dual duct constant volume conversion to cooling only variable air volume operation.

3 Constant volume reheat conversion to variable air volume with reheat operation.

4 Constant volume reheat conversion to variable air volume operation.

5 Multizone system conversion to variable air volume operation.

6 Unoccupied zone isolation.

7 Minimum outside air volume reduction.

8 Exhaust air volume reduction.

9 Exhaust air heat recovery.

10 Air system fan volume reduction.

11 Night setback operation during winter months.

## Water Side

1 Add a variable speed drive for pumps for hydronic systems with two way valves.

2 Costant flow hydronic system conversion to variable flow pumping operations.

3 Water balancing to individual terminal units.

## Major Mechanical Equipment

1 Add a smaller chiller for off peak load operation.

2 Add a heat recovery chiller.

3 Add a packaged boiler for domestic hot water.

4 Add flue gas heat recovery, economizers or combustion preheat systems.

5 Consider variable speed drive for chillers.

6 Consider variable speed drive for cooling tower fans.

## Automatic Temperature Control

1 Hot water temperature reset control for perimeter radiation.

2 Hot water temperature reset control for systems without perimeter radiation.

3 Chilled water temperature reset control.

4 Chiller sequencing.

5 Condenser water temperature control.

6 Space temperature set point decrease during winter.

7 Space temperature set point increase during summer.

8 Humidity control set point revision.

9 Economizer switchover revision.

10 Economizer set point revision.

11 Unoccupied period space temperature control reset.

12 Optimized equipment start and stop.

13 Supply air temperature control for all types of air systems.

## Electrical Considerations

1 Electrical demand limiting controls.

2 Duty cycling.

3 Power factor correction.

## Other Considerations

1 Janitorial scheduling.

2 Elevator, escalator scheduling.

3 Packaged units for computer facilities.

4 Solar reflective window film.

5 Exterior wall insulation (heat transfer reduction).

6 Exterior wall infiltration (air leakage reduction).

7 Domestic water temperature reduction and flow control.

## Heat Exchangers

| | | |
|---|---|---|
| 1 | Clean tubes. | AM |
| 2 | Check pressure relief valve. | M |
| 3 | Clean traps and strainers. | M |
| 4 | Record operating temperatures and pressures. | D |
| 5 | Check operation of controls. | D |

## Humidifiers

| | | |
|---|---|---|
| 1 | Remove drain plug and drain residue from bottom of sump tank. | W |
| 2 | Clean spray nozzle. | W |
| 3 | Check operation and calibration of humidistat. | W |
| 4 | Clean strainer screen in steam supply. | M |
| 5 | Check and clean steam trap. | M |

6 Check and clean water supply filter. M

7 Remove impeller cap and clean out pump tube. AM

8 Remove scale build up from unit. WN

## Incinerators

1 Lubricate motors. AM

2 Check combustion chamber lining. M

3 Clean out settling chamber. WN

## Plumbing

1 Check that fixtures and trim are mounted securely. W

2 Check flush valve adjustment. W

3 Ensure floor drain traps are water filled. W

4 Check washers, O rings and stem packings for leaks. M

5 Clean traps and strainers. WN

## Pumps

| | | |
|---|---|---|
| 1 | Check and record inlet and outlet pressure gauge readings. | D |
| 2 | Check packing for excess leakage. | D |
| 3 | Clean pump strainer screen. | M |
| 4 | Blow dust and dirt from motor windings. | AM |
| 5 | Clean pump housing. | WN |
| 6 | Lubricate pump and check bearings. | AM |

## Refrigeration Compressors

| | | |
|---|---|---|
| 1 | Check for oil or refrigerant leaks. | D |
| 2 | Check that the refrigerant charge is adequate. | D |
| 3 | Check that all controls are functioning. | D |
| 4 | Ensure that coil fins are straight. | WN |
| 5 | Check all electrical connections. | W |
| 6 | Check that the defrosting cycle is functioning. | D |
| 7 | Purge system if bubbles appear in the sight glass. | WN |

8 Clean coil, drain pan and drain the line. WN

## Unit Heaters, Cabinet Unit Heaters, Fan Coils and Force Flow Units

1 Check controls, thermostat, aquastat and PE switch. M

2 Clean the unit heating and or cooling coil. WN

3 Visually inspect fan operation and electrical connections. M

4 Clean filters or screens. WN

# AIR & WATER FLOW MEASUREMENT TECHNIQUES

## Air Flow

Air flow is measured in a building using various instruments and techniques. Below listed please find a brief description of the instruments and the techniques used.

## Duct Traverse

To measure air flow in a duct the duct traverse method is utilized using a thermal (hot wire) anemometer velocity measurement instrument. A straight run of duct must be located with at least 10 ft on each side of the test holes with no duct turns. If a rectangular duct, 3/8" test holes are drilled in the duct usually in the bottom or side. If a round duct, test holes are drilled in two axis at a right angle to each other. Multi air velocity test readings are taken over the cross sectional area of the duct and an average air velocity in fpm is arrived at.

This average velocity in fpm is multiplied by the cross sectional area of the duct in square ft which equals cubic feet per minute of air flow through the duct. Air flow is given in cubic ft per minute or litres per sec in the metric system. 2.12 cfm is equal to 1 L/s air flow. 197 fpm air velocity is equal to 1 meter per sec air velocity.

Thermal anemometers are available in two ranges, 0 to 3000 fpm and 0 to 10000 fpm. The 0 to 3000 fpm used for most duct systems.

## Static Pressure

To measure static pressure in a duct an electronic digital differential static pressure meter is used. There are also anolog magnehelic gauges available.

## Temperature and Relative Humidity

To measure temperature in a duct or a zone an electronic digital thermometer is used.

To measure relative humidity in a duct or a zone an electronic hygrometer is used.

## Supply Diffuser, Return and Exhaust Grilles

The ideal method to measure supply, return or exhaust air through a ceiling diffuser or grille is to use a collector hood. The collector hood is an instrument that measures air flow directly in cfm or L/s. The fumehood is complete with replaceable hoods of various sizes. The collector hood negates the use of calculations to determine the air flow.

If a collector hood cannot be used an electronic rotating vane anemometer or thermal anemometer may be used to measure air velocity through the diffuser or grille. Multi readings should be taken and then an average velocity be calculated.

In the case of a tiered supply diffuser the average velocity is multiplied by the free area (AK) of the diffuser to calculate air flow. The AK value can be supplied by the manufacturer or field calculated.

In the case of a return or exhaust grille the average velocity is multiplied by the size of the grille in square ft X the free area (AK) of the grille to calculate air flow.

## Water Flow

Water flow through pumps and other heating terminal units is usually done using circuit balance valves (CBVs) and computerized flow measuring devices.

For pumps, the CBV will be located at the discharge of the pump. For other terminal devices the CBV will be located in the return line.

The CBV serves three purposes, measuring water flow, balancing water flow and as an isolating valve. When the CBV is used as an isolating valve it is easy to return to the balance position after reopening the valve. The CBV is complete with a memory lock setting which allows the CBV to only be reopened to that preset balance position.

The computerized flow meter can be field programmed to measure flow for each particular CBV. The test probes for the flow meter are inserted into the test ports on the CBV. Using this differential pressure and noting the CBV stem open position the flow meter will measure and indicate the water flow through the CBV.

If a CBV is not installed it is still possible to measure water flow through a pump using the deadhead method.

Calibrated gauges must be installed at the inlet and outlet of the pump. There must not be a strainer or isolating valve located inside the gauge position.

With the pump at full flow note the suction and discharge pressures. Convert the resulting across pump differential pressure to ft head. 2.31 psi equals 1 ft head. Then close the discharge isolating valve and once again note the no flow differential pressure across the pump.

Using the manufacturer supplied pump curve, mark the no flow ft head pressure on the vertical of the pump curve. Using this point draw a pump curve paralell to the existing pump curve. Then take the full flow differential pressure and apply to the vertical of the pump curve. Using a straight edge note where this value intersects the drawn deadhead pump curve. Drop down to the bottom and note the water flow in usgpm.

www.ingramcontent.com/pod-product-compliance
Ingram Content Group UK Ltd.
Pitfield, Milton Keynes, MK11 3LW, UK
UKHW041821200726
13854UKWH00001BA/263

9 781425 948573